YOUR POCKET WATCH

THE OWNERS GUIDE TO MECHANICAL POCKET WATCHES

BY JOHN LOCKE

A BRABOURNE BOOK

Second edition. Published 2003
By Brabourne Books

brabournebooks@lockes.co.uk

Obtainable from www.lockes.co.uk

ISBN 0-9545590-0-2

TABLE OF CONTENTS

Introduction

Although pocket watches have been largely superseded by wrist watches, the latter are a comparatively recent introduction: only becoming popular during the First World War (1914-1918). Like mechanical clocks, well-maintained mechanical pocket watches will increase in value. They are works of art and miracles of small-scale engineering, and it seems amazing that the early ones could have been entirely hand made. The drawback is that each year there are fewer specialist watch repairers who are capable of maintaining and repairing them.

Timeline for Pocket watches

1400 — Mainspring

1500

Fusee invented

First portable timepieces worn on neck chain

Verge escapement. Foliot balance. Stackfreed regulation

English and French makers used fusee

Hog's bristle regulator. Worm set-up of mainspring

1600 — Case evolves from tambour or ball type

Striking and alarm watches. Glass fitted to case

Fob watch. Enamelling of case in colours

Pair cases

Spiral balance spring. Minute hand used. Tompion regulator. Going increased to 30hours

1700 — Enamelled dial. Chaffcutter escapement

Duplex escapement. Snap-in glasses

Cylinder escapement. Rack lever escapement

Chronograph escapement. Temperature compensation

Detached lever escapement. Self-winding. Tourbillon. Chronometer escapement

Improved rack lever

1800 — Maintaining power

Massey lever. Pump winder

Detached table roller

Keyless winding. No dust cap

Chronograph. Machine made watches

Karrusel

1900

Invar balance and elinvar balance spring

Types of Pocket Watches

There are several characteristics by which one could classify pocket watches. When the design of the case settled down, watch cases could be classified into three types: Open Face, Hunter and Half-Hunter

Open Face

Hunter

Half-Hunter

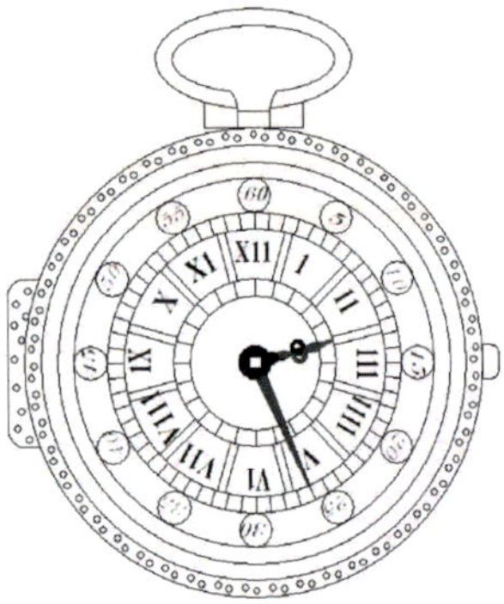

Paircase Watch

The paircase watch has two cases, the outer is often made of horn or shagreen and decorated with silver pinheads in the earlier models and of plain silver in the later ones. As can be seen from the illustration, the hinge of the outer case is at the side with a release button for opening it opposite.

However, apart from the type of case, the characteristic most obvious to the user is that some have to be wound using a key (keywind) and some have to be wound using the knob on top of the watch (stemwind).

Stemwind

Keywind

Within these types, are the various types of escapement used, including those mentioned in the timeline above.

How Pocket Watches Work

To use anything properly, it is necessary to have some understanding of how it works. This is for two reasons:

To avoid doing anything to the mechanism that is likely to damage it.

To understand the terms and to identify the parts involved in its operation and maintenance.

The Balance Wheel and Spring

In order to measure time, watches must run at a fixed rate. This means that, despite any variation of the power driving them, the hands must turn at a constant speed. To achieve this in mechanical watches requires a form of governor that only allows the power to run down (escape) in increments with a fixed period. This governor takes the form of a mechanical oscillator which is kept going by the escapement. In its later, accurate form, this oscillator took the form of a balance wheel controlled by a balance spring. A pendulum like those in clocks cannot be used because the watch has to operate in all positions.

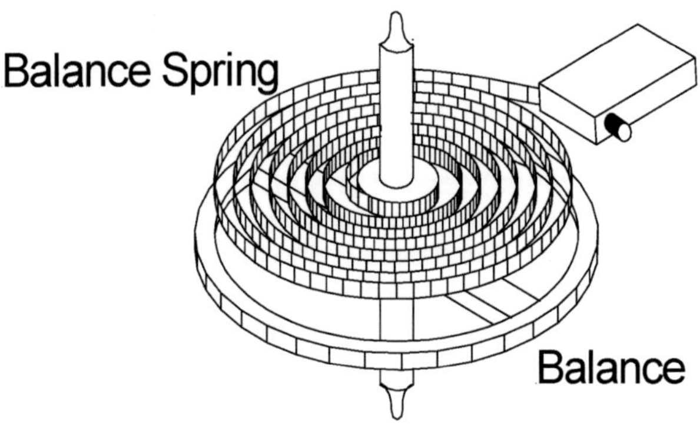

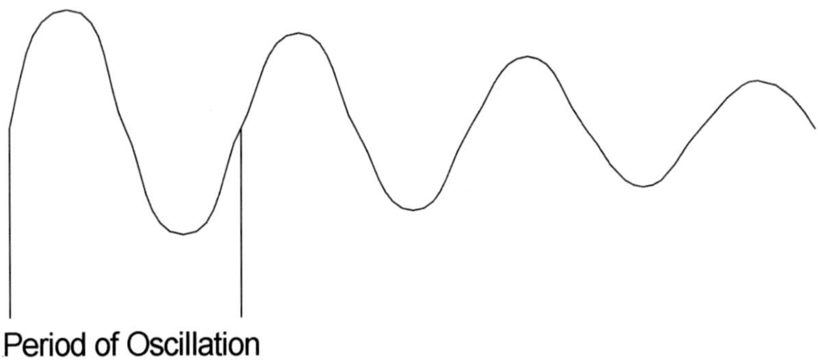

<u>Oscillations gradually dying away</u>

The balance wheel is free to turn on its axis. One end of the balance spring is fixed to this axis and the other end to the frame of the watch. Once the balance wheel is given a small amount of turn, the spring will be wound or unwound a little and will try to return the balance wheel to its central position. Being free to turn, the wheel will overshoot that central position and wind the spring in the opposite direction and so the process is repeated. Because of frictional and other losses, the oscillation thus started will gradually die away unless the balance wheel is given a small extra push each cycle to make up for these losses. The frequency of the oscillation (the number of cycles per minute) depend only on the moment of inertia of the wheel (made up of its mass and radius, both constant) and the stiffness of the spring. The device giving the small push each cycle is called the escapement and it is on this part that much research and development has been expended over the years.

Types of Balances

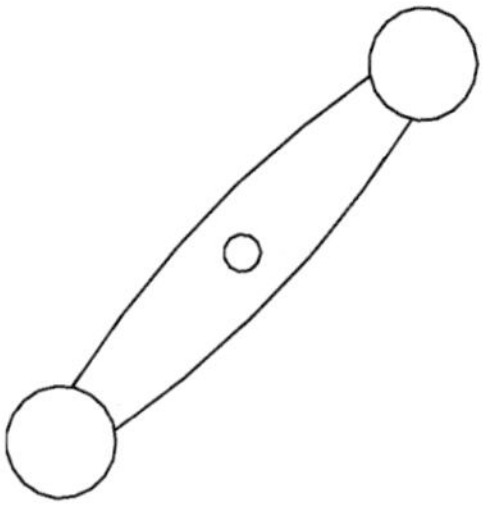

1560

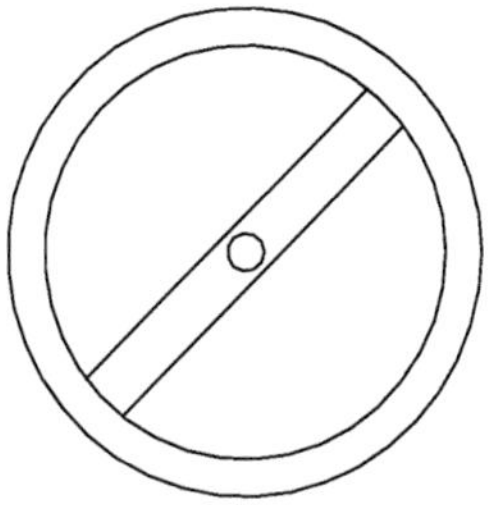

1600 →

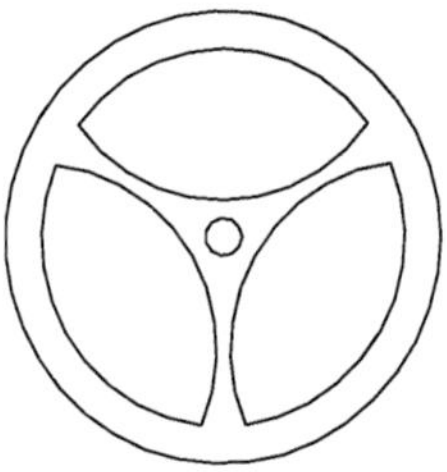

1640 →

1700-1900

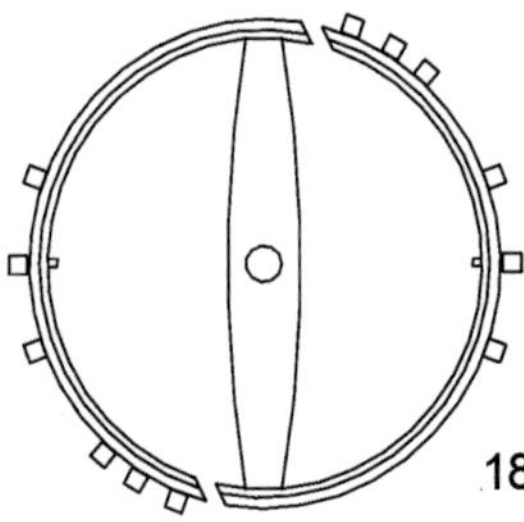

1800 →

The Escapement

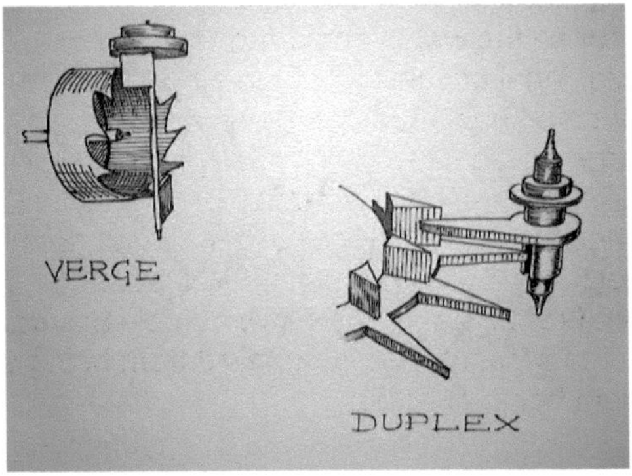

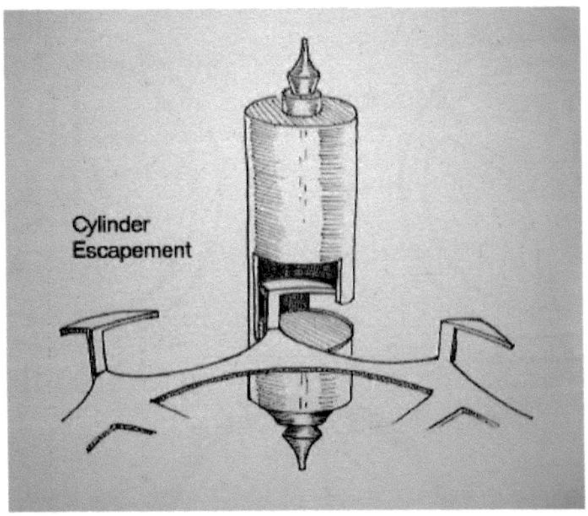

The earliest watch escapement was a verge driving a balance without any spring. The balance was just banged from side to side. The timekeeping of this device was not good, keeping time to about an hour or so a day. It depended a great deal on the power that drove it. The mainsprings that provided this power gave a driving force that depended on how much they were wound - fully wound, large force, nearly wound down, small force. Even with more advanced escapements, much ingenuity was expended to even out this force and devices evolved from a stackfreed through the fusee and mainspring set-up to the present going barrel. See the section on mainsprings.
Later, a hog's bristle was used as a rudimentary spring, but although it reduced the clatter of the verge on the balance, it did not do much to improve the timekeeping. When the spiral balance spring was

introduced, timekeeping was revolutionised, so much so that the minute hand was fitted and the train of wheels in the watch changed from three to four so the watch could run for 30 hours on one winding. An additional advantage of the spiral balance spring was that a regulator could be fitted to alter easily the timekeeping of the watch. The verge escapement reigned supreme for another 40 years when other types of escapement were introduced. The most successful of these was the cylinder escapement. Sixty years later, the acme of mechanical escapements, the chronometer escapement was invented and remained in use for extremely accurate timekeeping to the present day. It suffered from being expensive to produce and from being delicate.

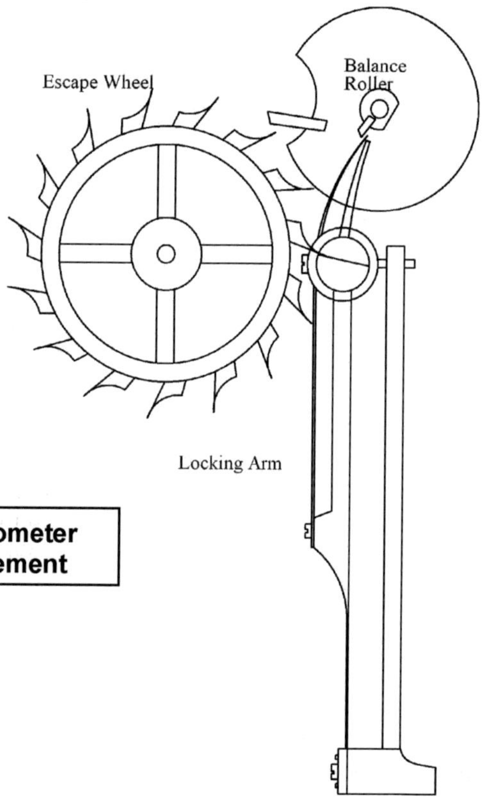

Chronometer Escapement

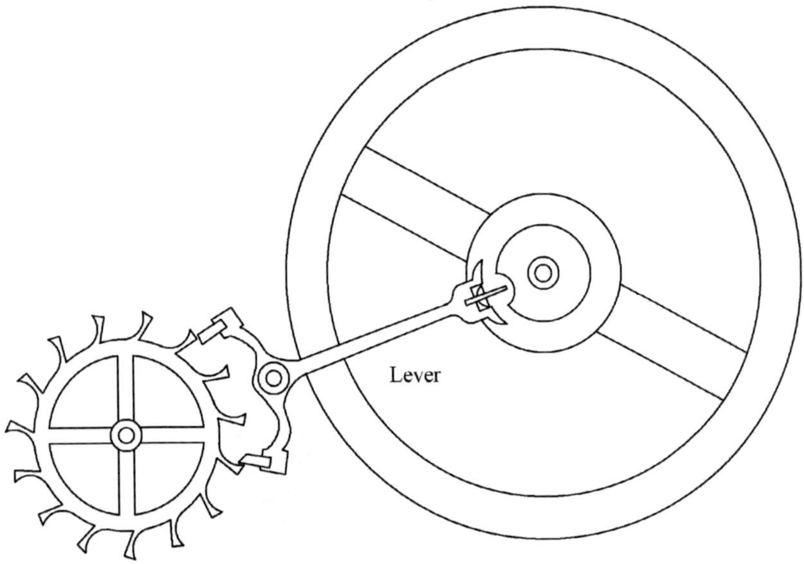

The Lever Escapement

Slightly earlier, the lever escapement was invented and with further development became and remains the standard escapement for everyday use. It is robust and accurate.

Because accurate escapements were now available, other factors that affected the timekeeping became noticeable, in particular, temperature effects and positional errors. The former, made the watch run slow when warm and fast when cold. Bimetallic compensation in various forms was used to correct this at first until metal alloys were produced which were unchanged by temperature changes (Invar and Elinvar). Positional errors meant that the watch would run at different speeds depending on whether it was on its face or side in various positions. One method of correcting for this was by slowly rotating the whole of the escapement while the watch was running (tourbillon or karrusel) and supplemented by careful shaping of the balance spring.

The Mainspring

The mainspring is a long strip of steel that is wound within a barrel. As mentioned above, the force exerted by the mainspring is not constant, i.e. it varies depending on whether the mainspring is tightly wound or nearly unwound.

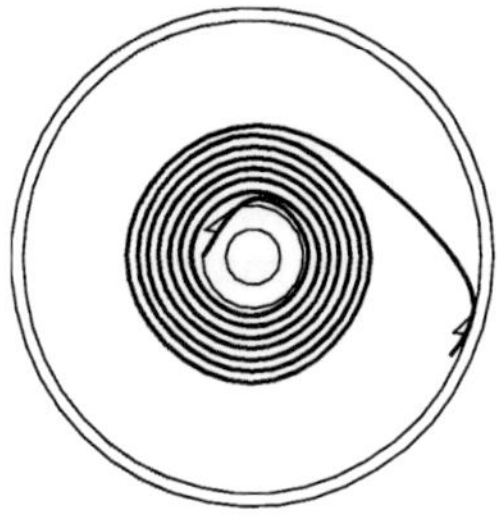

Mainspring in barrel, tightly wound

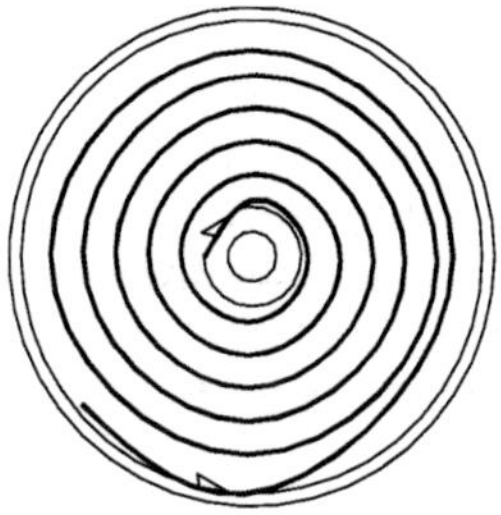

Mainspring in barrel, almost unwound

Over the years, various methods were devised to even out this mainspring force. They started with the stackfreed which was simply a friction cam exerting a large force to reduce the pull of the spring when fully wound and a small force when wound down. This wasted a lot of the mainspring's power and was one reason why early watches had to be wound twice a day.

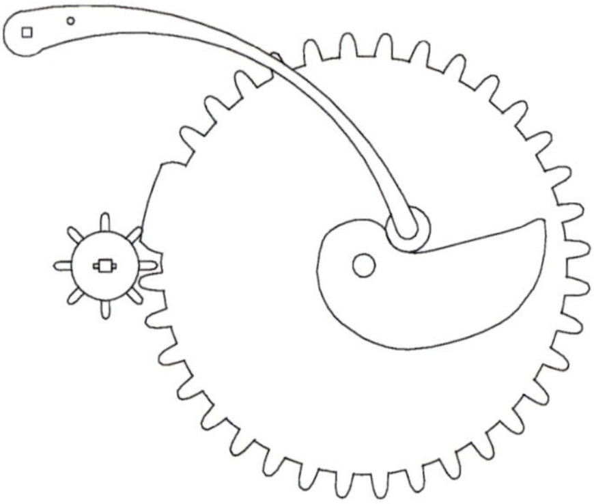

Stackfreed

A much-improved device was the fusee, which didn't waste power. The fusee is a variable diameter pulley between the mainspring and the rest of the watch movement. The fully wound mainspring operated on the smallest diameter of the fusee and the unwound spring on its largest diameter. The coupling between the mainspring barrel and the fusee is by gut line or chain. The fusee drives the rest of the movement by teeth on its periphery.

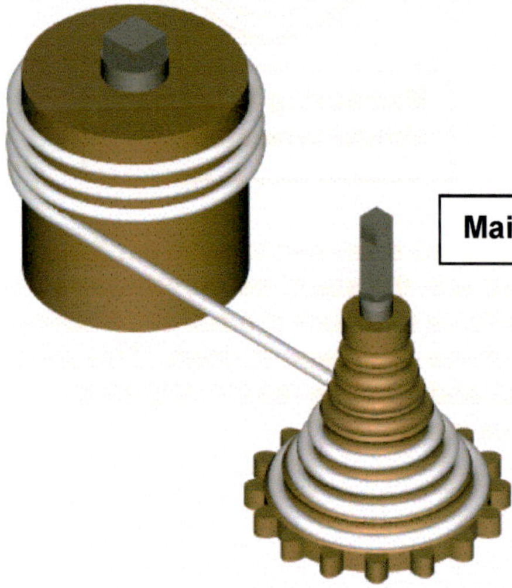

Mainspring Barrel & Fusee

After the adoption of the fusee, and when springs could be made longer, it was found that the force of the mainspring could be made more even by using only the middle section of its winding range. The initial amount of wind, to set the starting tension, is called the set-up. At first the set-up was controlled by a worm and wheel, turned by a very small key. This then evolved through a click set-up to Geneva set-up to no set-up today

Types of Setup

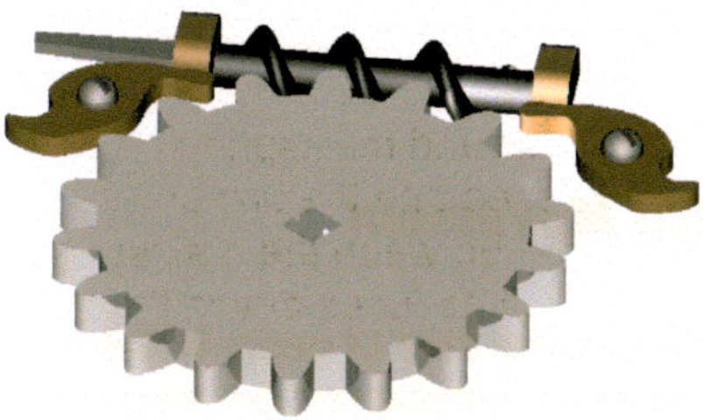

Worm Setup

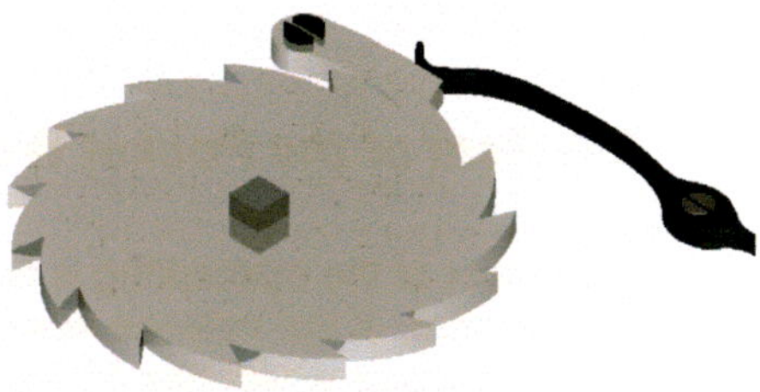

Click Setup

Geneva Setup

As mainspring manufacture developed, mainsprings became longer and their force more even between the wound and unwound states. The fusee could then be dispensed with and what is called a going barrel used. Teeth are cut on the edge of the going barrel to drive the rest of the movement.

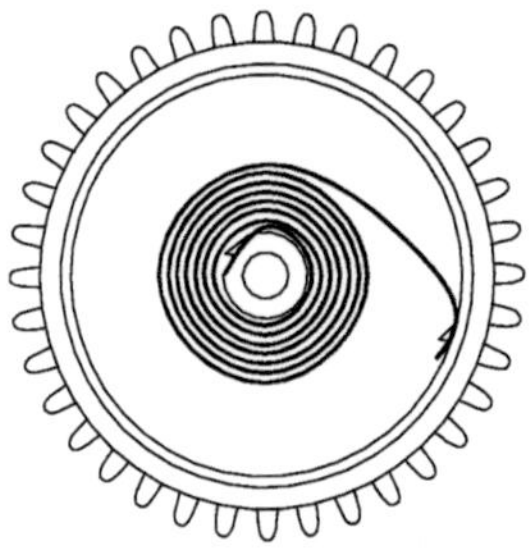

Going Barrel

The Rest of the Pocket Watch Movement

The power of the mainspring is fed through a train of gears to the escapement. If the balance wheel makes 5 oscillations per second and the escape wheel has 15 teeth, then the escape wheel turns 20 times a minute (one tooth per oscillation). The mainspring makes perhaps 8 turns in a day. If so, then the gearing between the two has to be 3,600 to 1. This requires four sets if gearing of a little less than 8 to 1. From mainspring to escape wheel, the gearing is always big to small.

In horology, the science of time, the big gears are called wheels and the small ones pinions. In all but the very oldest and rarest watches, the wheels are made of hardened brass and the pinions of hardened steel. An exception to this is sometimes found in the motion work, described later.

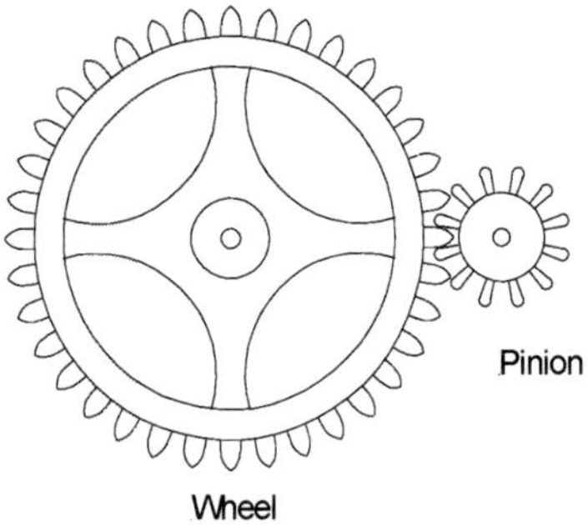

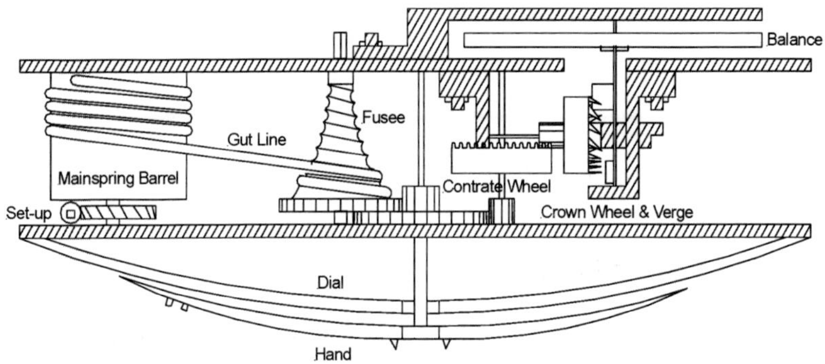

Early Verge Layout

The Motion Work

At some point in the movement, the hands must be fitted. If there are both hour and minute hands, then the minute hand is driven directly by the movement and the hour hand is driven from the minute hand by means of a set of wheels and pinions called the motion work. In the normal arrangement of showing the time, the minute hand makes one revolution per hour and the hour hand makes one revolution in 12 hours. The gearing between the two must be such that they both rotate in the same direction and about the same axis. An arrangement must also be made so that the minute and hour hands can be rotated independently of the rest of the movement when setting the watch to the correct time. This is obtained by making the motion work a friction fit to the main train of wheels and pinions.

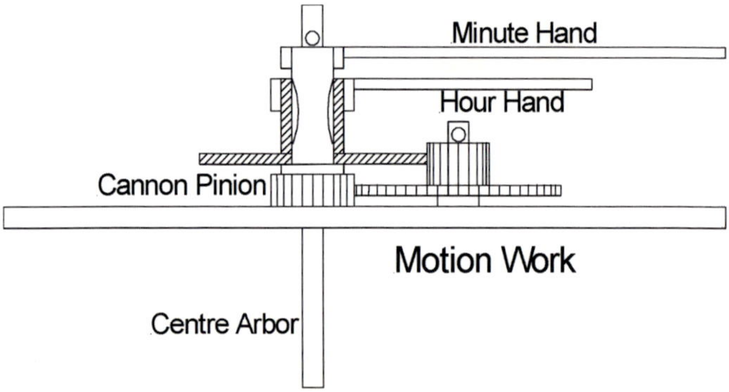

Operation

Opening the case

In order to see the time on a hunter watch, it is necessary to open the front of the case. With stem-wind watches, press the winding knob in while keeping your other fingers from interfering with the opening of the case.

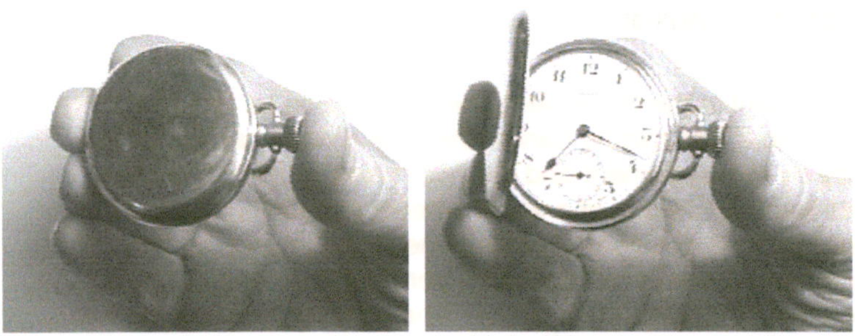

The same technique can be used with half-hunters to see the whole of the dial.

Key wind watches will have a either back cover which has to be opened to allow access to the winding square or the front glass has to be opened to allow access to the winding square in the dial. These are usually opened by pressing the button at the top of the pendant.

If there is no button, then the back will have to be opened carefully with a case knife (a stubby, blunt knife).

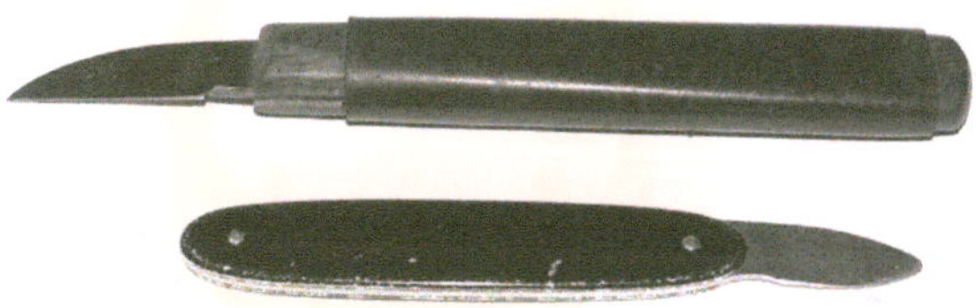

If a pair case needs opening for winding or setting the hands, there is a button on the side of the case that has to be pressed to undo it.

Setting the hands to time

Setting the Time on Later Stem Wind (Keyless) watches

Stem wind watches have several different methods of setting the hands to time. Most are set by pulling out the winding knob one click and then turning it to set the hands. Don't forget to push the knob back in after you have finished setting the hands. If you don't, the movement is likely to stop or keep poor time and the hands will

accidentally get turned. Many cheaper watches of the 'dollar' watch variety used a knob that was pushed in against a spring to set the hands and when released, the knob would spring out to the winding position.

Setting the Time on Earlier Keyless Watches

Earlier stem wind watches, used either an 'olive' near to the winding button which had to be pressed with a fingernail while turning the knob to set the hands, or there was a lever near the edge of the dial which had to be pulled out before the hands could be set. The front cover had to be opened to obtain access to this lever and the lever had to be pushed in before the cover could be closed.

Setting the Time on Key Wind Watches

Setting the time on these watches is accomplished either by opening the front cover (by fingernail on the projecting tab) and fitting the key onto the minute hand square and turning, or by opening the back cover and fitting the key onto the handset square and turning.

Winding

Winding Keyless Watches

Keyless watches are wound by turning the winding knob to the left. As all of this sort of watch are fitted with ratchets, so that if the winder is turned the wrong way it will not do any damage, the winder may be turned backwards and forwards to wind the watch. This will

still wind the watch on the forward turns. When winding, hold the winding button fairly loosely between the fingers so that when the watch is fully wound, the button will slip between the fingers. This avoids overwinding which can pull the end of the spring off the barrel.

Winding Key Wind Watches

Wound through back

Wound through front

To wind key wind watches, the key has to be inserted in the winding hole (in the back of most watches but in the dial of some), engaged on the winding square and turned. Many winding holes have an arrow to show which way to wind but, if you are not sure of the direction to turn it, try each direction gently. If the key turns easily in one direction then that is the one in which to wind it. DO NOT force the key or you will do expensive, or perhaps irreparable damage!. When winding is complete, a marked increase in the resistance to winding will be felt - don't wind any more or you may pull the end off the mainspring! Always make sure that you are using a well fitting key - it should not feel sloppy on the winding square.

Pocket Watch Faults

Pocket watches are comparatively fragile. Compared to a clock, their parts are small and the pivots very small and yet, for their size, the watches have a high mass because of the heavy cases - very often of gold or silver. This means that, even if dropped from a small height, the deceleration when they hit the floor or table is large and the pivots are likely to break. In fact, they will not withstand rough handling of any sort.

Mainsprings can break at any time. The metal of which they are made gradually becomes more crystalline with age and sudden changes of temperature may cause it to shatter. Over winding can, of course, break the end off a mainspring. A humid atmosphere will cause rust spots to develop on the spring and these become the focus of stress when the spring is wound. The spring can break at these points. Unfortunately, any spring breakage can result in other damage to the watch movement because of the sudden release of energy in the spring. The result that is most likely is bent or broken teeth on the mainspring barrel.

Mainsprings also get tired and end up by being unable to straighten out but remain curled fairly tightly in the barrel. The effect is that the watch will only go a few hours after being wound. The mainspring can easily be changed when the watch goes in for a service.

Few pocket watch cases are dustproof. Dust or fluff getting into the movement may stop it or even if it does not, the dust combines with the oil at the pivots to make a grinding paste which will wear the pivots away and increase the size of the pivot holes so that the pivots jam in them. The effect is much worse when the oil gets old and its volatile components have evaporated.

Old and sticky oil badly affects the timekeeping of the watch. In extreme cases, the whole movement seizes up.

Watch glass is easily broken, but is also easily replaced. The only trouble is that when the watch was made, the glasses were mass-produced and were cheap and every watch repairer kept a large stock of all sizes. Such days have now long gone and each glass has to be made especially by hand and is expensive.

Maintenance

As can be seen from the above, watches should be regularly serviced. The recommended period is every two years whether the watch is being used or not. This involves stripping down the movement to its component parts, cleaning out the old oil and dust, polishing pivots and pivot holes, checking the mainspring and replacing it if necessary, reassembly, lubricating, testing and adjusting.

Complicated Pocket Watches

Not all pocket watches just tell the time, Since the very first pocket watches, additional complications such alarm or moon phase or other astronomical indications have been added. Some types are described below:

The clock watch

This early type of watch sounds the hours on a bell.

The alarm or appointment watch

This is fitted with an extra hand or hands which can be set to the time that it is desired that an alarm is to be set off.

The Repeater Watch

These watches sound the time on demand by pushing and releasing the pendant or a lever at the side of the watch. The usual types are quarter repeaters, and minute repeaters. The time is usually sounded on two gong wires that encircle the movement but remain clear of the case so that, when struck by miniature hammers, they give a clear bell-like sound.

A repeater watch by LeRoy.
The gongs can be clearly seen reflected in the back cover.

The striking is in ting-tang form where the ting is a high note and the tang is about a third lower. The time 5.47 would be sounded on a quarter repeater as ting-tang, ting-tang, ting-tang, ting-tang, ting-tang, ting, ting, ting i.e. the last quarter was at 45 minutes. Some repeaters were designed as 'dumb' repeaters, without gongs with the hammers striking directly on the case so that only the wearer would be aware of the vibrations.

The Multi-Dial Watch

This type of watch shows not only the time but also the month, day of week, date and often the phase of the moon. These additions are changed automatically by the main movement and can also be set by pushers around the rim of the case.

Chronograph Watches

These have an extra time interval measuring section fitted which is engaged or disengaged at will from the main movement without interfering with its time-keeping properties. The chronograph section is started, stopped and zeroed by pressing on the pendant. Some chronographs have a split second display with a separate pusher for starting and stopping one second hand without interfering with the other. Chronographs have the usual time display with a subsidiary seconds dial. The time interval measurement is generally displayed by a long centre second hand and an elapsed minutes subsidiary dial.

Note the difference between chronograph and chronometer. The former has been described above whereas the latter is a high quality timekeeper which has gained a certificate for its accuracy. Chronometers often have a special escapement to ensure this accurate timekeeping.

Examples of European Pocket Watch Movements

Verge by Asby & Hutchins, London 1730

Verge by William Haywood, London 1722

Verge by
Robert Jones,
Ruthven 1800

Verge by
Bradshaw & Riley,
Coventry1810

Fusee Lever by
S. Broadhurst,
1820

Fusee lever by
Brownhill,
Liverpool ca
1825

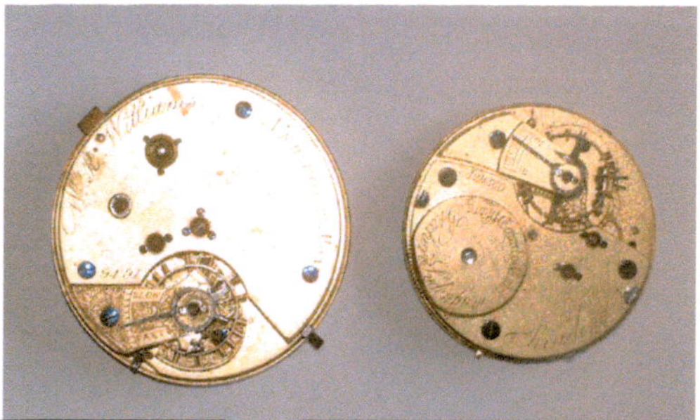

3/4 plate fusee lever by W.R.Williams, Newport

3/4 plate fusee lever by Bracher & Sydenham, Reading ca 1877

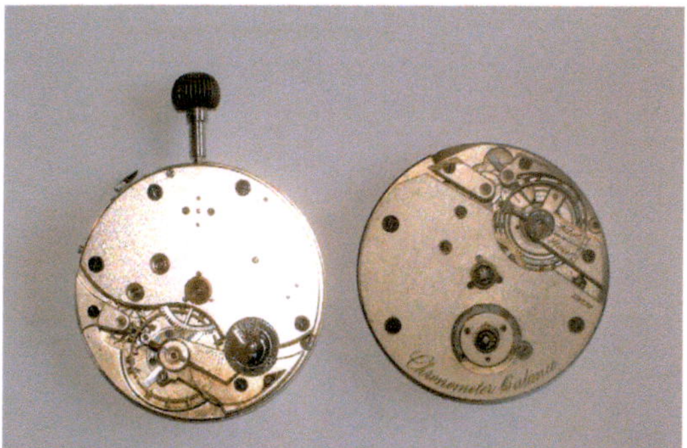

High quality Swiss 3/4 plate with cam regulator, ca 1900

run-of-the-mill Swiss 3/4 plate going barrel lever

Swiss 3/4 plate chronograph 1887

Swiss Geneva bar chronograph ca 1900

Going barrel lever by Reid & Sons, Newcastle

Good quality Geneva bar lever

Silver fusee verge by J.Stedman, London 1775 showing how movement hinges in case

fusee lever by Henry Mortimer, Liverpool

Examples of American Watch Movements

Elgin Watches

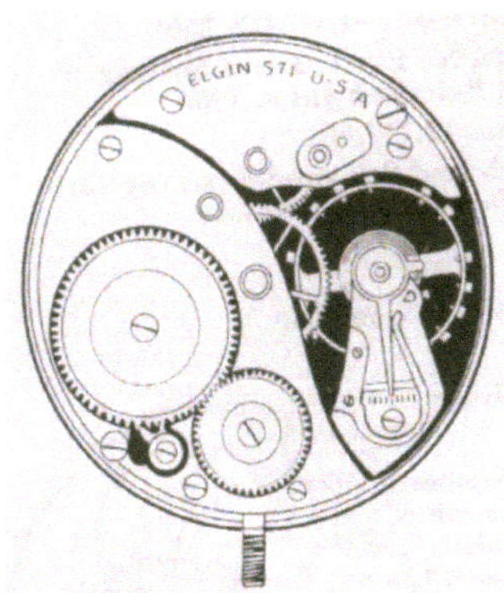

Hamilton Watches

GRADE 992B

GRADE 950B

Illinois Watches

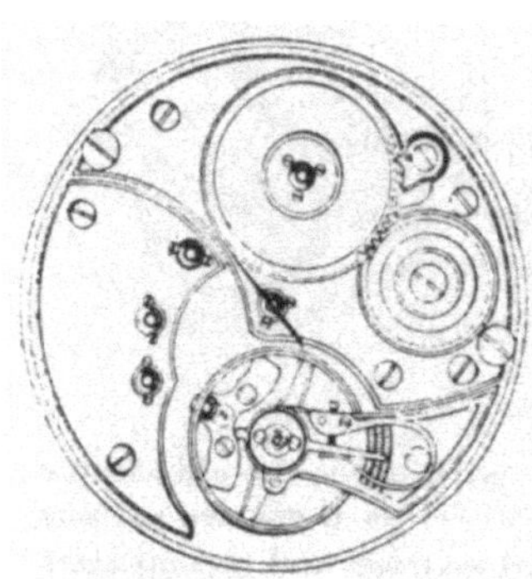

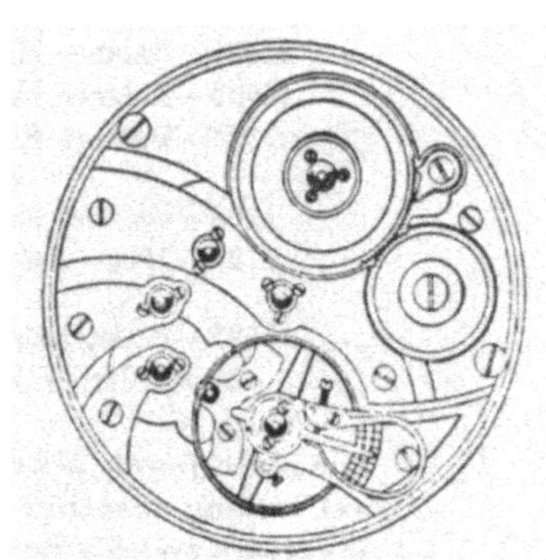

Waltham Watches

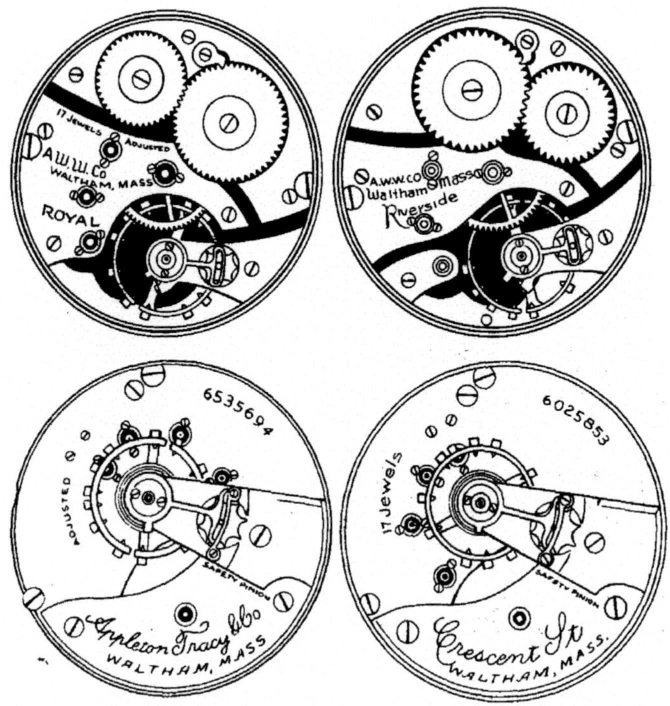

HUNTING OPEN FACE

New York Standard Watches

EIGHTEEN SIZE

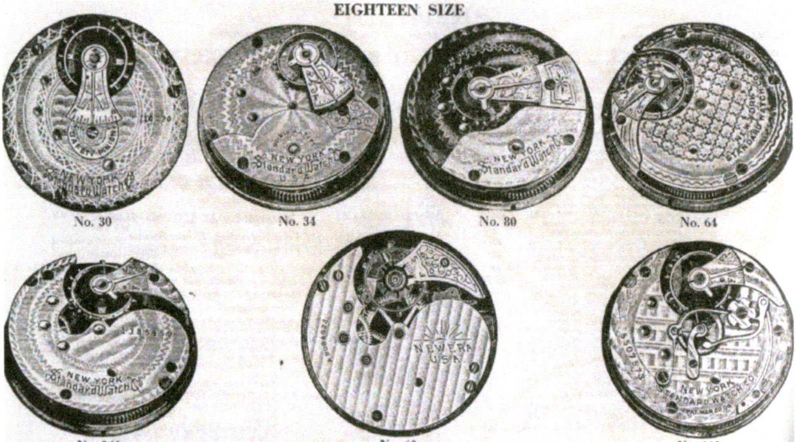

No. 30 | No. 34 | No. 80 | No. 64

No. 360 | No. 60 | No. 165

INDEX